When Buffy did chores, she got a dime, a penny, and a pack of jelly candy.

Buffy said, "I don't want games or pretty stuff. I just want Yum-Yums!"

Yum-Yums came in funny shapes.
Buffy had a bunny, a kitty, and more.
"The best shape is the puppy," she said.

At the store, Buffy put in a penny to get a candy. Which candy did she want? A Yum-Yum!

"The best Yum-Yum I see is a bee," said Buffy. "I'll do chores. Then I'll put in a penny and get it."

Buffy fixed up her room and made it look pretty. She made Sandy look pretty too. After that, she would help her mom.

At the store Buffy put in penny after penny. With her last penny she got the bumblebee!

What was this? Not a jelly candy!
It was a plastic bee that buzzed.
"What a funny bee!" said Buffy.